I0478431

The Day My Smartphone Ended!

The Trials and Tribulations of an over-60 Smartphone Owner

by

Stephen M Kraemer

Introduction

As an over-60 smartphone owner, I have learned a lot about phones and technology in the last few years. At the same time, I have encountered many problems and issues with these incredible devices. I have written this book in order to share these "smartphone adventures" with other owners and users, so they can appreciate and hopefully benefit from my experiences.

Trial 1

Buying the Phone

(<u>And </u>a Few Other Things to Go with It!)

"Thank you," I said to the dealer after buying my phone. "Thank you very much." I was happy now to have my new phone, which of course I would have to pay a certain amount for every month in order to keep. "You mean there are more items I have to buy in addition to the phone?" I asked the dealer.

"Yes," he said smiling. "You'll need a cover for your phone and a screen protector, and you'll need an extra charging cable."

"Of course," I agreed. "Of course." Naturally, I did not want to spend any more money than I was already spending for the phone itself, but I could see where a case would be quite beneficial. Since I'd be carrying the phone around and using the phone all day, a case would be very important, particularly since I tended to be pretty rough in handling devices in general. A screen protector also seemed like a good idea.

I certainly did not want to have to worry about smudges or scratches on the screen!

"Okay," I said. "I'll buy the case." The dealer was very helpful, and actually attached the case to the phone, which was not an easy task.

I was very grateful that the dealer did this for me, so I would not have to struggle with trying to do this myself at home.

"And I'll go ahead and get the screen protector also," I said to the dealer, feeling very pleased with myself that I was now adding the right accoutrements to my new device.

"Well," he said, I actually do not have the correct size screen protector here in stock, so if you want to buy one, you'll have to go to the mall or to another store to get one."

I was not too thrilled at the prospect of making another trip to acquire a screen protector, but I figured that I had gone this far getting the device ready to use, so I might as well finish the task.

"Oh, and by the way, how much will this screen protector cost?" I asked before leaving the store.

"About forty dollars," said the dealer, "if you get a glass protector, which is what I recommend."

"Forty dollars?" I asked in a louder voice. "You mean I need to spend another forty dollars just to put a cover on top of the screen?"

"That's right," said the dealer. "That's the price."

I thanked the dealer for all of his help – he had been very helpful – and proceeded to the mall to get a forty-dollar glass screen protector for my new smartphone.

Trial 2

Recharging

About two weeks after getting my new phone, I woke up one morning and promptly checked my smartphone to see if the battery registered 100% after recharging the phone all night.

"What? – 35%?" I yelled almost in shock! "That can't be. It's been charging all night. Why isn't it at 100% now?"

I calmed down a bit and realized that I needed to relax and just think carefully about what had happened.

There must have been some reason why the battery in the phone did not recharge overnight, and if I thought about it, I would probably find the answer. Soon after, when I checked the charge cord carefully, I could see that the connection of the cord in the charging socket was not very good. Obviously, I had not had a good connection overnight, so the phone could not charge up. I fixed the connection of the cord in the socket, and then I looked at the phone to see if the little icon at the top was indicating that the phone was charging, and it was!

I let it go ahead and charge up for the rest of the morning, and hopefully I would have a fully charged phone in a relatively short period of time. By the time I actually got ready to use my phone a couple of hours later, it was fully charged.

Congratulations to me on my success!

It was an easy fix, but if I had not discovered it, I would have started worrying about whether the battery was working properly, and what other remedial actions I would have had to take!

Trial 3

Light at Night

I went to bed one night with my new smartphone charging, and noticed that I was having problems falling asleep.

"Why can't I sleep?" I asked myself. Something is bothering me – something is keeping me awake! I noticed there was a little blue light in the corner of my room.

"A blue light?" I murmured. "I don't have any blue lights in my room."

I always turn off all the lights in my room before going to bed, as I do not like any lights on at night. Even the smallest light on in my room would keep me awake, so I always make sure that all the lights are completely off before going to sleep.

Still, this blue light continued to bother me, and I couldn't understand where it was coming from, at least not until I walked over toward the phone. And there it was – a tiny but powerful blue light emanating from my phone!

Suddenly I had to figure out what to do. The phone was plugged in and recharging, and I always recharged my phone at night while I was sleeping. But I certainly could not keep the phone recharging now, not if this blue light was going to keep me awake all night.

Naturally, I took action! I unplugged the charging cable from the socket, and decided to just let the phone stay uncharged overnight. It wouldn't be fully charged when I woke up the next morning, but at least I would get some sleep. I wouldn't have that little blue light bothering me all night long! I just had to remember to plug the charge cord back into the socket when I woke up the next morning, and recharge the phone early, so I'd be able to use it the next day. That's all I had to remember to do!

Trial 4

Lost Cord

I was on a cruise ship traveling to Alaska in September when I took out my phone to recharge it overnight. Naturally I looked for my charge cord, but guess what? It wasn't there! And, without the cord, I couldn't charge my phone. I started to think:

"Where was the charge cord? Was it in one of my suitcases? Did I leave it somewhere?" I had to think of where I had been recently. Where was the last place I used the cord to recharge my phone, and could I have left the cord there? Then it dawned on me! I had stayed in a hotel room the night before, and I must have left the charge cord in the room. This was the first time I had checked out of a hotel without the charge cord for my cell phone!

The first thing I thought of now was how to charge up my phone so I could use it. I went to the gift shop on the cruise ship to see if I could buy another cord for the phone. I did not see any cords in the gift shop, so there I was with my phone, but no way to charge it! Luckily, I ran into someone at the photographer's studio on the ship, and the person in charge listened to my story and offered to charge up my phone for me, using one of the phone chargers at the studio. I would have to leave the phone at the studio for a while, but that was okay with me. I really did not need my phone very much on the ship anyway, so leaving it was not a problem. I was very happy that I was able to find someplace to charge the phone, and thanked the photographer at the studio several times! He had saved my cell phone, at least for a while.

My next task was to obtain a replacement cord for the phone as soon as possible, so that I would be able to continue charging the phone every day. As I could not purchase a cord on board, I waited until the ship docked in Juneau, and proceeded to buy a cord at one of the stores in town. I bought a cord (it wasn't cheap of course), and also had to buy a plug for the cord, since a cord by itself, without the plug for the socket, is useless!

Now, with both cord and plug in hand, I was very excited, and could not wait to get back to the ship to try it out. I did so later, and Eureka, it worked! I could actually charge my phone now. I had accomplished a very important step in the process of keeping my phone going!

By the way, I should mention that when I got back home, I could not use the plug that I had bought for the charge cord, since it would not fit properly into the wall socket for recharging. I had to replace it with another plug!

Another interesting note: I called the hotel where I thought I had left the original charge cord to see if it might still be there. I spoke to someone in housekeeping and they went to my room and looked for it. They not only found it, but were willing to ship it to me. I thought this was very nice of the hotel, but l decided against it, since the cost of the postage would have been more than the cost of replacing the cord myself!

Trial 5

Contacts

One day (as I am sure many have experienced), a note appeared on my phone requesting a "system update." This seemed a reasonable thing to do, since an updated system should be better than an un-updated one. I went ahead and accepted the update, and waited for the new system to arrive. When I received the update, I noticed some changes in different parts of the phone operation. Most noticeable were the changes in the email system, which now showed emails more clearly, and this was a very nice change.

Soon afterwards, I went to the Contacts section of my phone to make a phone call, and sure enough, my contacts were gone! I found this hard to believe, but it was undeniable: I had no contacts in my phonel! This obviously had been the result of the recent system update, so what was I supposed to do?

The answer came when I went to see the smartphone dealer.

The system update had been on my new phone, a second phone which I had purchased after my previous smartphone had gotten old. After some discussion with the dealer, he suggested that I install a special backup program on both my old and new phones. Even though my old phone no longer had calling features, the rest of the phone was still working, so all of my original contacts should still be saved on the old phone. By using this backup program on both phones, it should be possible to transfer all of my contacts from the old phone to the new one.

We proceeded to try this out, but it was not easy. First, I had to install the backup program on both phones. Then, I had to log in to both phones using the same username and password (both phones had the same operating system). The dealer and I then went through a series of steps, backing up the contacts on the old phone, and then sending them over the Internet to the new phone. This took quite a while, and there were problems, but eventually it worked. I now had all my contacts on my new phone.

Sometime later (as you might have guessed), there was another system update on my new phone, and what do you think? Yep – my contacts vanished again! But luckily, I had backed them up this time, so I used the backup program that I had saved on my phone and was able to retrieve all my contacts.

This was wonderful! I had learned and used all these technical tricks to save my contacts, and I was thrilled at my accomplishments.

By the way, so you don't think I became too smug about this, the contacts did not last forever. Eventually, the special backup program failed, and the contacts disappeared again. This time I simply decided to enter any new contacts into my phone by hand, and not worry about having the full list of original contacts in my phone.

I thought to myself, "Do I really need all of those original contacts in my phone? Many of them are quite old, and I don't even use most of them anymore. I'll be fine with just the current contacts that I have, and that will be enough!"

Of course, eventually – that's right – I'll have to back them up!!

Trial 6

When a Phone Is Not a Phone

It was a nice fall day in October, and I had just turned my phone on in the morning, when I saw something circling on the screen. It was a little arrow going round and round on the screen, and there was a note that said, "Updating 31 of 91 apps."

I thought to myself, "What does that mean? And where is my phone?" I could not see the normal start screen for the phone. All I could see was this update arrow going around in circles. It was taking several minutes for this update, and I assumed it was the computer part of the phone that was updating. But it was obvious that I could not use my phone during this update! I would have to wait until the update was finished before I would have access to my phone. This meant that for at least several minutes

I HAD NO PHONE!!

Then I began thinking, "I cannot use my phone right now, and I won't be able to use it for at least several more minutes, not until this computer update finishes."

Then I began to experience a minor panic. "What if there's an emergency in the next few minutes?" I thought. "How will I be able to make a phone call if something happens? And what if there's another update at a later time when I need to make an important call?"

After a few minutes, the update finished, and I had my phone back again. But the sense of panic did not go away. I began to get very worried about what might happen in the future. If my phone's computer decided it wanted to update during a true emergency, what would I do? I wouldn't have a phone, and I wouldn't even have an button for 911, because I would have absolutely no access whatsoever to my phone during the update! I'd be in an urgent situation without a phone!

I then realized that the only solution would be to secure a second phone. I had no landline at home, and no other backup phone, so I was completely dependent on my current cell phone for contact with the outside world!

After a little more deliberation, I decided to go to my smartphone dealer, and see if I could purchase a second phone. I went to the dealer, and was offered several options, including buying a phone with a relatively inexpensive plan – a type of pay-as-you-go phone. This would be affordable, but I would have to keep buying minutes every few months to keep the phone going. Otherwise, I would run into lapses on the phone, and would not be able to use it until I had added the additional minutes needed. I decided that this option was not for me. If I was going to get a second phone, I wanted something that would be available all the time - a phone where I wouldn't have to worry about adding minutes or having a computer update happen at any time. The answer was simple – I needed a flip phone – and not a smartphone!

The dealer then offered to give me a flip phone for free – an extra phone that he had sitting around. I thought this was very nice of him, and I accepted the offer. Of course, there would still be the expense of owning and operating a second phone! This meant spending a certain amount of money every month for a second line, and getting a second phone number for the new phone. I was not too happy about spending more money, and adding to my phone bill, but I was more concerned with having a backup phone for emergencies. I was happy with the result, and I could finally relax.

Trial 7

Over and Out

Winter was here, and I picked up my smartphone one morning and turned it on. At least I thought I was turning it on! I pressed the button on the side of the phone for the standard five seconds that it usually takes to turn the phone on, but it didn't turn on! "Okay," I thought to myself, "just try again, and hold it a little bit longer. It might take more than five seconds to turn it on today."

I held the side button on the phone longer, and started counting seconds, so I would know how long it would take for the phone to turn on. I held the button, and held it, and kept holding it for almost twenty seconds!! This may not seem like a long time, but when you are holding the button on the side of a smartphone for fifteen seconds longer than you should be, this is a long time – a very long time! The phone finally came on, and worked well, as it usually did, but I began to get concerned.

"Would this happen again?" I wondered. This was certainly not normal, and probably was a sign that something was wrong with the phone.

Sure enough, a few weeks later, it happened again: a full twenty seconds to hold down the start-up button to turn on the phone! Before long, I went back to the smartphone dealer, and mentioned this issue to him. He put the phone through a diagnostic and checked it out, and could not find anything wrong. This was good news, but I was still concerned. I was not used to taking twenty seconds to turn on my phone, and what if one day it took longer than twenty seconds to turn on, or did not turn on at all? What would I do then?

The dealer was very sympathetic, but made it clear that my phone was getting old, and would probably have to be replaced before too long. This phone did not have a replaceable battery, so once it stopped working, the only alternative would be to purchase a new phone. I was not too happy with this news, since I really liked my phone! I had become quite attached to my phone – its size and weight, the way it handled email, the apps I had on it, and everything else about it. It would be a shame to have to get a new phone. But I would have to resign myself to the fact that sooner or later I would have to give it up, and make a new purchase. I would have to realize that smartphones simply do not last forever.

And let's not forget – my smartphone was almost three years old – a virtual lifetime in the smartphone world!!

www.ingramcontent.com/pod-product-compliance
Lightning Source LLC
Chambersburg PA
CBHW061225180526
45170CB00003B/1158